The Spilled Coffee Chronicles™

The Portion PadL™

Equal Slice Pizza Cutting Board

Volume 1 of Several Volumes

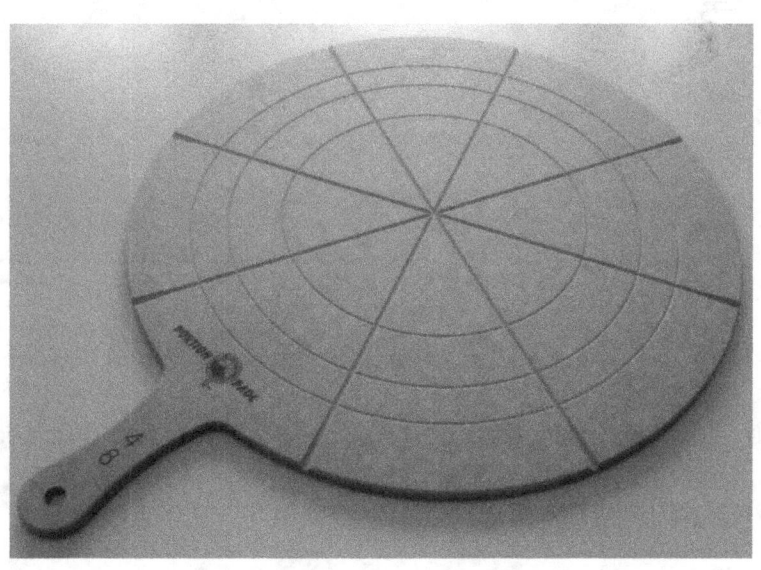

Copyright 2010 Spilled Coffee Chronicles™, Owner Andrew R. Spriegel

The Pizza Cutting Board

First edition

Copyright 2010 Spilled Coffee Chronicles™ and Andrew R. Spriegel

Author: Laura Freeman

Contributing Authors: Andrew R. Spriegel and Greg Getzinger

Edited by Michael Massaro

Visit our website for additional information and products at:

www.PortionPadL.com

Table of Contents

Introduction
This is a workbook, use it like one!

The Spilled Coffee Chronicles^{tm} is a series of volumes that document real world examples of an actual invention or business. This series (the Equal Slice Cutting Board) is made up of numerous volumes. The Equal Slice Cutting Board series is the factual story and the actual experience of inventor Greg Getzinger and Andrew Spriegel, a patent attorney, as they worked together to move an invention from an idea to a successful, commercialized product.

Greg's invention could have remained in his pizza shop as merely a personal tool for his own convenience, but he showed his invention to Andrew. Andrew was impressed by the simplicity and elegance of the board design and he drew on his background as an inventor and engineer to add important new/novel features to the board.

Andrew added his expertise in patent law, management and manufacturing with Greg's experience to sell and market their invention. This is how they did it, and you can do it too with your idea, learning from their successes, failures, strategies and determination.

> **DEFINITION –** *According the Webster's dictionary, an inventor is someone who produces (as in something useful) for the first time through the use of the imagination or of ingenious thinking and experiment.*

Each individual invention is covered in numerous volumes of the Spilled Coffee Chronicles™. The various volumes focus on different aspects of each invention and the principles which Andrew or others used to bring the invention to market. By following the steps provided here, you can increase your odds of taking your invention from an idea to a commercialized product. You will learn from both the successes and failures encountered during the process. With multiple volumes it is easy for a potential inventor to choose the volumes that they need. For example, if the inventor needs help with marketing they can purchase the volume on marketing and ignore the volumes where they have strengths, for example, design.

In this first volume, we will explore how Andrew met Greg, a pizza business owner who had an idea for a better pizza cutting board. Together, they developed the Portion PadL™ into a viable profitable product in only a few months.

OTHER VOLUMES: WILL COVER MARKETING IN DETAIL, SALES, MANUFACTURING, PRODUCTION; MANAGING CASH FLOW, INVENTORY CONTROL, DETERMINING PRICING, WEBSITE DEVELOPMENT, ETC.

Why was the Spilled Coffee Chronicles™ series created? Because Andrew Spriegel, a patent attorney, had an electro-mechanical engineering and management background for developing products for other companies prior to going to Law School. He created the Spilled Coffee Chronicles™ to help inventors commercialize their products without falling into the potholes Andrew and other inventors have fallen into during the invention process.

Andrew started his career in engineering. After graduating from the Rochester Institute of Technology (RIT) in 1986, he worked on a wide range of products such as locomotives, off-highway vehicles, satellites, computers, surgical products, medical products, pumps and consumer products. Just like many other engineers or employees, Andrew made a lot of money for other companies with the products he developed. In many cases, he was either the sole inventor or a co-inventor of the products.

After working for several companies GE, IBM, Lockheed Martin, Invacare and others, Andrew grew tired of being dependent on employers for work and not sharing in the profits resulting from his inventions.

FACT – *According to the Bureau of Labor Statistics (www.bls.gov) people born between 1957 and 1964 will hold an average of 10.8 jobs from ages 18 to 42.”* ***That is approximately 2.2 years per job!***

After many years of dependence on employers, he went back to school at the University of Akron Law School to become a patent attorney with his wife Beth's full support, love and help. He often says Beth should have gotten half of the diploma for all the support and encouragement she gave him during law school.

> **FACT** – *Recent statistics from the U.S. Department of Education show that adult students are the fastest growing educational group and these numbers are steadily increasing. The Institute for Higher Education Policy reports that **students age 40 and older increased by 235 percent from 1970 to 1993.***

New York Times – By Abby Ellin, Published: November 11, 2006
Photo: Emile Wamsteker for The New York Times
"As Older Students Return to Classrooms an Industry Develops"

A common thread often among inventors and entrepreneurs is they want ownership in their efforts, a share in the profits and the fruits of their labors and to be able to control their own destiny. They are often tired of worrying if they will be part of a layoff! Many experts are predicting that the current economic recovery will be a "jobless recovery" and many people are now forced to become entrepreneurial or try to commercialize their ideas.

QUOTE

"Adversity has the effect of eliciting talents which in prosperous circumstances would have lain dormant." - Horace

What are the pros and cons for you if you make a voluntary career change?

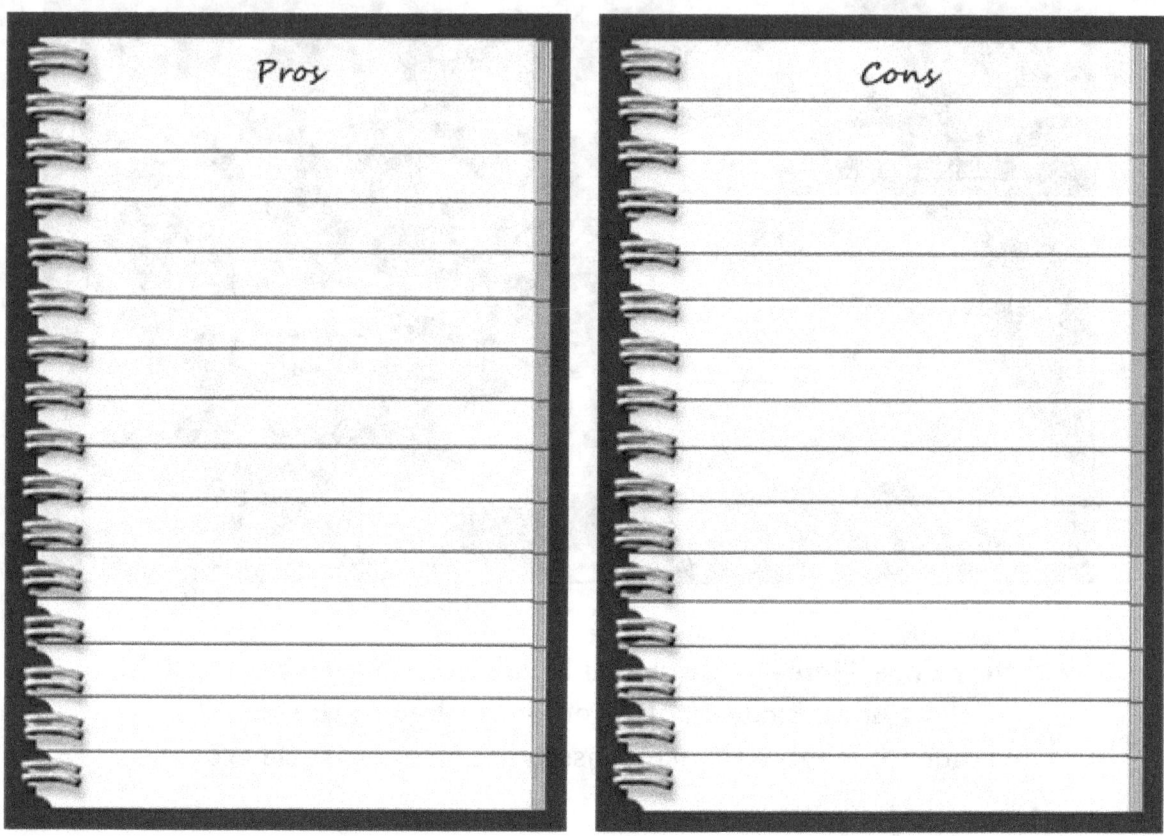

Fill in the notebook, it will help with completing a Detailed Business Plan, by completing the notebooks in the various Volumes of the Portion PadLᵗᵐ

PATENT AND TRADEMARK ATTORNEYS

www.Smart2Patent.com

Andrew learned to take a product from start to finish during his engineering/management career and has applied this knowledge in working with inventors. In addition, he has contacts with companies, designers, engineers, CNC (computer numerical controlled) machine operators, tool makers, computer aided designers, prototype builders, marketing and sales professionals, manufacturers and others.

The Spilled Coffee Chronicles[tm] documents actual inventions and the efforts to commercialize them. As mentioned previously, each invention is written as a series of volumes, where each volume concentrates on an aspect of that invention and the principles used to bring the invention to market. The Spilled Coffee Chronicles[tm] of Invention for the Pizza Cutting Board has numerous volumes: a volume about building a prototype, a volume on marketing the product, a volume on building a website, *etc.* By following the steps provided in each volume, you can increase your odds of successfully taking your invention from an idea to a commercialized product. You will learn from both the successes and failures that you will read about in the volumes. It is often said that we learn more from our failures than from our successes.

Pizza Cutting Board
Volume 1, Chapter 1
Identifying a Problem: Different Size Pizza Slices

The irony of the situation wasn't lost on him. As a police officer, Greg Getzinger had never been asked to go undercover. But now, years after leaving the force, he was at a high school basketball game, undercover, seeking information.

He had positioned himself just a few feet from the concession stand. Close enough so that he could overhear the customers' comments, but not so close that he would draw attention to himself. He smiled pleasantly and leaned against the wall trying to look as if he was enjoying the game.

But the information he wanted wasn't for a criminal case. It wouldn't put anyone behind bars. In fact, no one else even cared.

A steady stream of customers had been buying candy, popcorn and sodas since he arrived, but he wasn't interested in any of that. Greg was there because of the pizza. More specifically, he wanted to find out what customers *thought* about his pizza.

That is because Greg is the owner of pizzaBOGO, a small pizza business in Hudson, Ohio. PizzaBOGO is a gourmet; carry out award winning pizza business. But Greg wasn't always an independent businessman.

www.pizzaBOGO.com

After graduating from college, he became a police officer, married his high school sweetheart, Alisa, and settled into a career in criminal justice. But after three years he left the force to become an insurance agent. He found that doing police work was not as exciting as he imagined and he wanted to do something in sales or business. He started at the bottom but eventually rose to vice president of a big insurance agency.

However, after 17 years in the insurance business, he felt dissatisfied and unfulfilled. After a lot of soul searching, he quit the insurance business and used his life savings to buy a pizza business. Many of his friends and family thought he had lost his mind or was going through a mid-life crisis, but Alisa, believed in him and together, they made the plunge.

According to a study reported by Beth Parker of Fox TV in 2010, "only **45%** of those polled are satisfied with their jobs. That number was **61%** in 1987.

Are you satisfied with your job or are you with the majority of people that don't like their job? Why have so many people started new businesses? Why are you reading this volume of the Spilled Coffee Chronicles of Invention?

Some of your reasons may include fear of a layoff, job dissatisfaction or you lost your job and you have heard that this is going to be a jobless recovery. Is it a desire to control your own destiny, the ability to keep the fruits of your own labors and/or the fulfillment of a personal dream?

Pizza Cutting Board
Volume 1, Chapter 2
Overcoming Fear

FEAR

The brave man is not he who does not feel afraid, but he who conquers that fear. - Nelson Mandela

Fear often prevents people from becoming successful entrepreneurs. Many people are afraid of living their dreams paralyzed by fears. Fear is often not based on reality and it prevents actions and pursuit of goals. Many well-known and successful organizations were born during an economic slump including The Hyatt Corporation, CNN, MTV, Sports Illustrated, LexisNexis, Burger King, Microsoft, FedEx and General Electric to name a few.

Think about what you want to achieve by starting your own business or commercializing an idea/invention. Voice fears to a trusted friend, spouse or associate rather than acquaintances. Those close to you will tend to give more honest advice. Find a support network within your community, a business association, inventor groups, SCORE (a volunteer group, Counselors to America's Small Business), *etc.* However, don't be surprised if your friends doubt or discourage you from going out on your own or pursing your dreams.

Remember the key is to just keep moving forward to build your confidence. Sometimes just moving forward a small amount at a time can help keep you on track enough to reach your goals. Just starting the process can create enough momentum to move you down the path toward success. Action will eventually produce the results that you need to pursue your goals.

Entrepreneurship and the goal of owning your own business is a powerful driving force and the same is true with inventing. Many fear a personal lack of knowledge. While not everyone is an expert in public relations, sales and marketing, accounting, law, production and distribution, these seemingly lofty skills can be learned or contracted. Books and Internet research resources are FREE at the public library and courses are readily available that teach hands-on practical skills. Lack of knowledge and information is only limited by your imagination. You may even discover that you know more about business than you had previously thought possible. **WARNING:** In addition to good advice there is bad advice books out there (*i.e.*, those authors/books with no credentials and those with poor success rates).

Entrepreneur Stories:

Tastefully Simple

Can do Attitude, it MATTERS!

Jill Blashack Strahan started her gourmet food company, Tastefully Simple, with around $6,000 in savings, a backyard shed for storage, and a pool table as a packing station. She said: "I remember sitting outside one day, thinking we were **three months behind on our house payment**, I had two employees I couldn't pay, and I ought to get a real job. But then I thought, no, this is your dream. Recommit and get to work."

She wanted to have products sold at taste-testing parties. In 12 years, Tastefully Simple has grown into a $120 million dollar business.

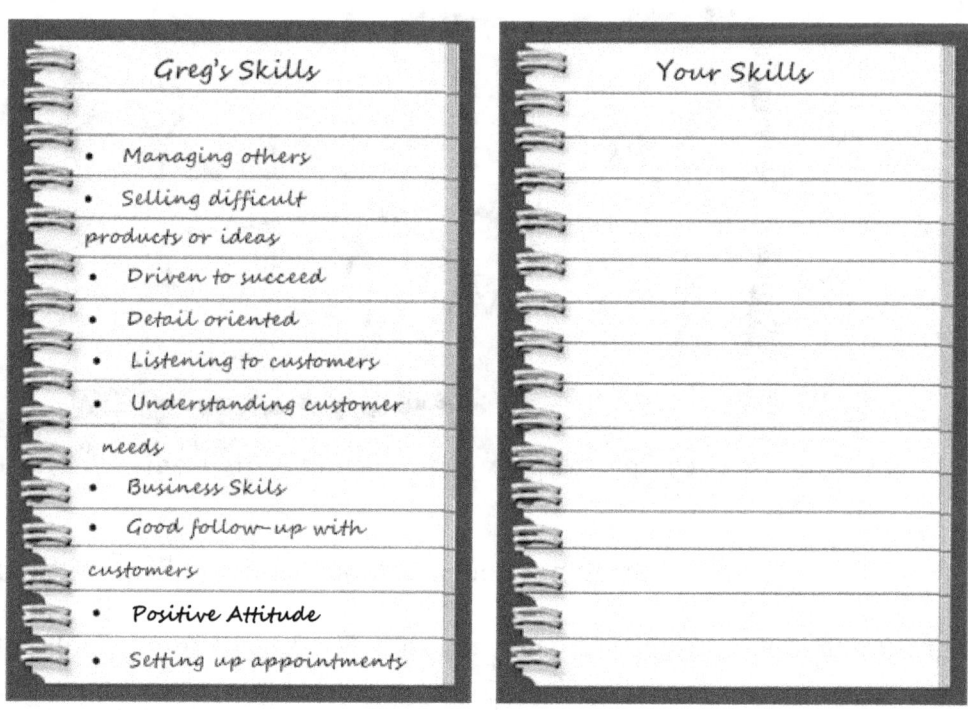

QUOTE

"The more intensely we feel about an idea or a goal, the more assuredly the idea, buried deep in our subconscious, will direct us along the path to its fulfillment." - Earl Nightingale

Greg's Skills	Your Skills
• Managing others • Selling difficult products or ideas • Driven to succeed • Detail oriented • Listening to customers • Understanding customer needs • Business Skills • Good follow-up with customers • Positive Attitude • Setting up appointments	

Fill in the notebook

To Greg the pizza business was an "on-the-job" learning experience. He purchased a franchise in an existing business that had been around about a year and had a small customer base. Greg began to develop the business with a successful coupons and specials program, but he wanted an even larger customer base. Greg discovered that sales were not growing the way he wanted. The pizza shop was not a large franchise and coupons, though helpful, were not effectively promoting his product very well against larger, well known pizza shops that had more money to advertise.

"I wanted to compete with larger pizza franchises and get into the schools," Greg said. "Once kids tasted my pizza and recognized who sold it, they would tell their parents." Greg said the feedback he received confirmed his theory. One of the main reasons that parents came to his shop, pizzaBOGO, was because their kids were eating it at school and telling their parents how much they liked the pizza.

And Greg was determined to make it a success — to prove all the naysayers wrong. He worked long hours at the business, learning everything he could about the pizza business.

He learned what ingredients to buy, how to store them, how to mix them, how to make the pizzas, and most importantly, how to sell them.

He **worked hard** to find ways to increase sales and he did well. In the first six months, he met his sales goals and his prospects looked good. But like any good business owner, Greg wanted to do better. His quest had taken him in many unexpected directions. One of those directions was supplying pizza to schools, which resold them by the slice in the cafeteria and at sporting events. And now he found himself in a high school gymnasium trying to gauge customers' reactions to the pizza that his business had supplied to the concession stand.

Throughout the game, each time someone bought a slice of pizza, he asked the customer for comments. As he expected, the responses were favorable.

After an hour, sales slowed and Greg decided to call it a night. But as he was about to leave, the woman running the concession stand asked if he wanted anything. On a whim, he decided to buy a slice of his own pizza. Nothing like first-hand product research, he thought. What happened next surprised him and launched him on a journey that would change his life.

"Do you want a big slice or a small one," the woman behind the counter asked.

Greg was confused. The pizzas had been baked and cut into individual slices back at his shop. All the slices were the same — there were no 'big' or 'small' ones.

"What do you mean," he asked. "Aren't they all the same?"

"Every pizza has one or two pieces that are smaller than the rest," she said. "And no one wants the small ones."

She then said, "It doesn't matter which pizza place comes in here, they're all the same. No one can cut equal pizza slices."

"What do you do with the small ones?"

"We usually have to throw them away at the end of the night."

He hadn't given much thought to the size of the slices. When a whole pizza is sold to an individual, no one cares if some slices are slightly bigger or smaller. But he immediately saw the problem when pizzas were to be sold by the slice. Every slice had to be the same size or someone would likely feel cheated.

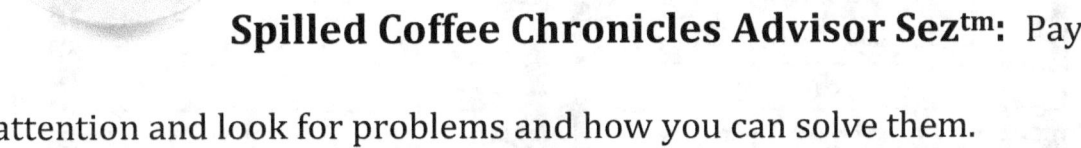 **Spilled Coffee Chronicles Advisor Sez^{tm}:** Pay attention and look for problems and how you can solve them.

If you want to invent something pay attention to the problems you and others encounter each day and think of ideas or products that can solve those problems.

Inventions are often the result of identifying a problem and coming up with a solution, but sometimes inventions come about by accident. For example, Alexander Fleming **discovered the drug penicillin by accident**. While organizing a pile of containers where he had been growing bacteria, he opened each dish and examined it before dropping it into the cleaning mixture in the sink. A particular container caught his attention. Mold was growing on one of the containers, but in another container the mold stopped the bacteria's growth, killing it. He named it penicillin and found that it was nontoxic and efficient in treating many types of bacteria harmful to man.

Dr. Fleming used his power of observation to identify something that was unusual. Greg found there was a problem with the size of the pizza slices by observing. If he wasn't paying attention he would have missed the problem.

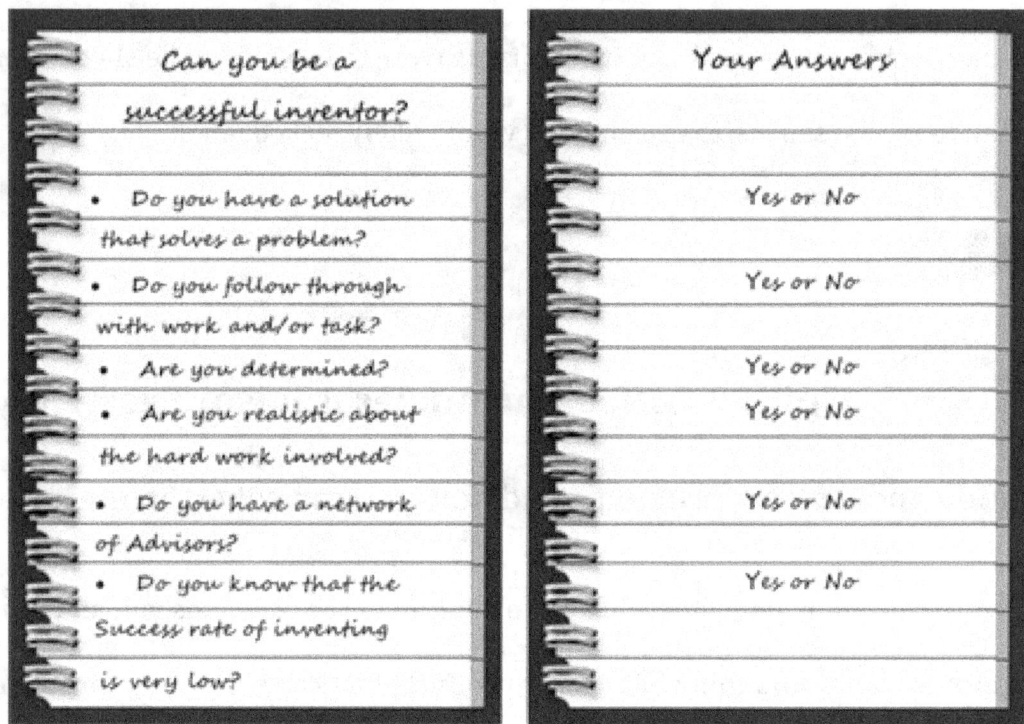

Fill in the notebook, it will help with completing a Detailed Business Plan, by completing the notebooks in the various Volumes of the Portion PadL™

Inventors need to know if they have what it takes to develop their idea into a viable product. Inventing is not for the faint of heart and it is not easy, if it was, all inventors would be successful. They need to know their strengths and weaknesses. Greg had an invention, but he had to take a hard look at the skills he had for producing and marketing his invention. **<u>Understand that people have gone bankrupt, gotten divorced, spent their life savings, *etc.* on inventing.</u>** You really need to assess your skills and drive to invent before you try to commercialize a product.

"Leonardo da Vinci's Workshop: Inventor, Artist, Dreamer"

exhibit in New York

"Too many inventors think if they come up with an invention, people will automatically want to buy it, but that's the wrong idea. Commercializing products is hard to do, and the inventor turned businessman needs to be driven. If they are not hard workers, they won't be successful at inventing. In addition, the inventor is often the wrong person to market the product," Andrew says. Inventors often want to talk about how their product works and not how it will benefit the customer.

However, this was not the case with Greg; he had a strong background in sales and marketing.

Andrew spends time evaluating clients that are inventors to find out who has the **tenacity**, **skills** and **ability** to work with others and stick with the process of commercializing a product as well as the right product to commercialize.

"I'm looking for someone who will go through all the muck, the missteps, the heartache, the ups, downs, changes and hard work to get their invention to a certain level where someone will buy it; and there are not a lot of people like that," Andrew said. "I've found there are groups of people who have backgrounds that help them to be successful at inventing — those who served in the military, those who are already businessmen, those who are successful in their career, and those with a 'can do' attitude."

Commercializing an invention requires a lot of money, a large amount of time, the ability to withstand the mental ups and downs, interpersonal skills and a strong desire to be successful.

The United States is a "first to invent country," whereas most other countries in the world that have a patent system are "first to patent countries." Every inventor should keep a record of his inventing work. A science lab student keeps a notebook of problems, theories, testing and solutions. An inventor needs to keep a record of all written materials and a back-up of computer information about ideas, designs, materials, testing, research and solutions, too.

Not only does the notebook provide insight about the inventions progress, it can provide proof of when the inventor conceived the idea.

ANOTHER VOLUME: In another volume you will learn the proper way to keep and format an invention notebook and why you need to keep a patent notebook.

Inventor Factors that Help with Success

Here is a list of factors/skills that we consider important in inventing; check the skills that you think you have in order to be an inventor. Place a check next to Yes or No.

Yes		No		Good at observation
Yes		No		Entrepreneurial (want to run your own business)
Yes		No		Hard worker
Yes		No		Have financial resources
Yes		No		Marketing Skills
Yes		No		Design skills
Yes		No		Sales skills

Yes		No	Determination
Yes		No	Passion for your work
Yes		No	Understand that new businesses or inventions have a high probability of failure
Yes		No	Can build prototypes
Yes		No	Can you assemble a team that can overcome your limitations?
Yes		No	Business management skills
Yes		No	Computer skills
Yes		No	Website development skills
Yes		No	Higher education
Yes		No	You set goals
Yes		No	Prior entrepreneurial experience
Yes		No	Relevant industry experience
Yes		No	High level of drive
Yes		No	High level of energy
Yes		No	Able to establish a clear vision
Yes		No	Enough self-confidence to take carefully calculated risks

Yes		No		Committed to success
Yes		No		Have perseverance
Yes		No		High tolerance for uncertainty
Yes		No		Believe that you can control your own destiny
Yes		No		Ability to get other people to work with and for you
Yes		No		Know what motivates you and other people
Yes		No		Readiness to learn from your own mistakes and failures
Yes		No		Network personally and professionally
Yes		No		Do you network with successful people?
Yes		No		Does your family/husband/wife support your venture?
Yes		No		Do you network with people that support your venture?
Yes		No		Does your product or service have a market?
Yes		No		Do you have a business plan?
Yes		No		Do you have successful mentors?
Yes		No		Do you have successful associates?
Yes		No		Do you have a board of advisors?

Yes		No		Is your product patentable?

Fill in the Checklist

Add up the number of Yes answers you checked: _____

The more yes answers you have the more entrepreneurial you are and if following the proper strategy you will have a higher likelihood of success. Greg scored a 33 on this chart.

Here is a board of advisors that Greg has to increase his likelihood of success:

Accountant	His Dad
Day to day support	His Wife Alisa
Someone to manage the Pizza Business	Will Shaw – General Manager
Patent Attorney	Andrew Spriegel
Bank	Morgan Bank – Local Community Bank
Bookkeeper	His Dad
Manufacturer 1st Composite Board	Ohio Amish Woodworkers/ Manufacturers
Manufacturer 2nd Composite Board	West Coast Manufacturer
Manufacturer 3rd Composite Board	Massachusetts Manufacturer
Laser Etching	Ohio Laser Etching Company
Product Liability Insurance	John Fink – Insurance Agent
Marketing Support	Greg Getzinger

Sales Support	Greg Getzinger
Order Fulfillment (Shipping & Handling)	Manufacturer
Packaging	Manufacturer
Video Production	Andrew and Greg

Spilled Coffee Chronicles Advisor Sez™: It is very difficult to commercialize products on your own without help and advisors.

Who are Your Advisors?

Accountant	
Day to day support	
Patent Attorney	
Bank	
Bookkeeper	
Manufacturer	
Prototype Builder	
Material Supplier	
Product Liability Insurance	
Marketing Support	
Sales Support	
Order Fulfillment (Shipping & Handling)	
Packaging	
Video Production	

Fill in the table

Pizza Cutting Board

Volume 1, Chapter 3

Is there a solution to the problem?

At first Greg didn't think anything could be done to guarantee equal size slices. Didn't all pizza businesses have the same problem? Why should his pizza business be any different?

But the equal size pizza slice problem wouldn't go away. Greg sold pizzas to fundraisers, which operated on slim profit margins. Like the schools, they bought whole pizzas at one price and sold them by the slice at a higher price. Unlike other merchandise, such as candy, pizza cannot be stored. Pizza that is ordered to sell at a concession stand has to be sold, taken home or thrown out at the end of the night. Unsold slices cut into the fundraisers' profits.

Wasted pizza hurt Greg's business reputation as well. Greg wanted to keep his customers happy and that meant helping them reduce waste and increase profits.

Unequal Slices – No one wants the smaller pieces

Greg had identified the problem and he began searching for a solution. He had numerous conversations with the general manager of his pizza business, Will Shaw. (Will has helped Greg throughout the invention process by running the pizza business while Greg marketed the Portion PadL^{tm}.) Together they tried to solve the different size slice problem. They tried placing marks on a standard cutting board and practiced cutting perfectly equal slices. But no matter how hard they tried, they couldn't create equal slices consistently.

The problem was compounded by the fact that pizzas come in different diameters and must be cut into different numbers of slices. Some pizzas had to be cut into six slices while others had to be cut into eight, ten or twelve slices.

In addition, Greg recognized that if the pizza wasn't properly centered on the board the slices would not be equal. The board had to accommodate different diameter pizzas from the smallest to largest. It also had to be easy to use, food-safe and reasonably priced.

After a lot of searching, Greg realized there was no product available on the market that solved all of the issues and/or problems in a novel way. The few products that attempted to solve this problem and were on the market, were difficult to store and more important, difficult to use.

Then, he had an **idea**. Why not cut grooves into a pizza cutting board that would guide the knife and include circles for centering the pizza? Why not put grooves and circles on both sides of the board to cut different size slices and make the board food safe and easy to clean. It seemed like a simple idea, yet nothing

with all these features was available in the marketplace.

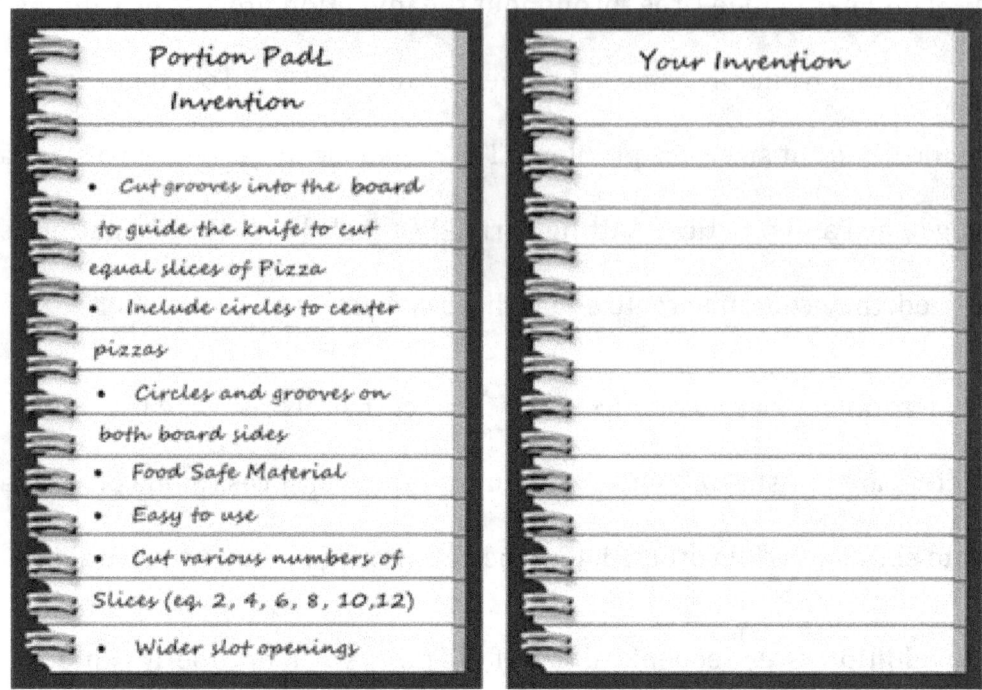

Fill in the notebook

Pizza Cutting Board

Volume 1, Chapter 4

Is there a need for this product?

Is there a need for this product? If there isn't a need there is no reason to

go any further. If there is a need then what does the product have to do?

Spilled Coffee Chronicles Advisor Sez™:

Face reality, you need to determine if people will buy your product.

FACT – Inventors are human — they love their ideas but often fail to determine if there is an actual need for the product. Many inventors spend tens of thousands of dollars on their invention only to find out no one needs or wants their product.

ANOTHER VOLUME: Although we briefly discuss how to determine if there is a market for a product, in another volume, we will describe in much greater detail how to determine if there is a need for your product, how to market your product and how to price your product.

Greg spent a large amount of time researching the advantages of the equal slice pizza cutting board. One day, during one of Greg and Andrew's road trips to the Ohio production facility, Greg stopped at several pizza establishments to pick up a to-go menu and inquire about their pizza sizes and number of slices. It was on this trip Greg made a big discovery. All the pizza establishments that Greg visited sold different size pizzas and different numbers of pizza slices. Yet all the current cutting devices on the market were all sold as a standard design. Greg realized that the cutting boards would need to be custom made based upon the buyer's specific pizza applications, in terms of the number of pizza slices and the diameters of the various pizzas.

Greg's conclusions about a custom made equal slice cutting board:

- Custom made means perfect pizza slices for each customer.

- Equal pizza slices eliminates the smaller, thrown out pizza slices which increases business profits.

- An easy to use custom made cutting board shifts employee responsibilities. Inexperienced employees can be assigned to cut the pizzas into equal slices. The more experienced employees are free to work on the more skilled projects which speeds up production time and reduces labor cost.

- Local, State and Federal governments are requiring "Nutritional Facts" labeling placed on food items. The ingredients that go into making a pizza are measured with measuring devices like cups, ladles and scales, yet when pizzas are cut unequally, the nutritional values are no longer balanced. A custom made, equal slice pizza cutting board assures that the "Nutritional Facts" are accurate.

Nutrition Facts

Serving Size
Servings Per Container

Amount Per Serving

Calories   Calories from Fat

% **Daily Value***

Total Fat	%
Saturated Fat	%
Trans Fat	
Cholesterol	%
Sodium	%
Total Carbohydrate	%
Dietary Fiber	%
Sugars	
Protein	

Vitamin A   •   Vitamin C
Calcium   •   Iron

* Percent Daily Values are based on a 2,000 calorie diet. Your Daily Values may be higher or lower depending on your calorie needs:

	Calories:	2,000	2,500
Total Fat	Less than	65g	80g
Sat Fat	Less than	20g	25g
Cholesterol	Less than	300mg	300mg
Sodium	Less than	2,400mg	2,400mg
Total Carbohydrate		300g	375g
Dietary Fiber		25g	30g

Calories per gram:
Fat 9   *   Carbohydrate 4   *   Protein 4

FDA Nutritional Facts

- Custom made centering circles makes the cutting board easy for the employees to use. **The cutting device must be "user friendly" or the employees will not use it!**

There were other products on the market that attempted to solve the same problem. Having other products on the market that endeavored to cut equal slices validates that people recognize the problem and the fact that the other products are marketed and sold indicates there is a market for the product. However, all of the previous products suffered from numerous issues, a few of which are discussed for each product.

He needed to make sure the targeted industry would have a demand that would support the sales of the invention. The Portion PadL^{tm} would be targeted toward commercial pizza franchises, large food service companies, medium and small businesses and the consumer market.

Targeted Domino's Pizza

Targeted Schwan's a Large Food Service Business

However, all of the products on the market wouldn't work well for all the various pizza applications Greg needed for his business. The cutting device he was looking for needed to be something his employees would want to use, especially during the busy lunch and dinner hours. It also had to be easy to clean. In addition, it had to be easy to store and financially affordable.

When Greg researched the pizza cutting devices, he noticed that there were numerous issues with each of them. Much of Greg's research came from personally visiting local pizza establishments, convenience stores, school systems, bowling alleys and the Columbus Ohio Pizza and Ice Cream Convention to learn from the employees and managers the strengths and weaknesses of the cutting devices they were using. All were open about their experiences and observations, a few of which are mentioned in the following pages.

Spilled Coffee Chronicles Advisor Sez™:

Thoroughly research the products that will compete with your invention. Evaluate each one and compare them to your invention. How is your invention different and novel? Why is your invention different? Why will people buy your invention over what's currently being sold?

Shown below is a pizza cutter guide. Though it performs as promoted, cuts equal slices, it has issues that make the device difficult for the end user, the people who matter the most. They need to cut equal pizza slices with ease, especially during the busy, high volume lunch and dinner hours. For example, the pizza cutter guide isn't easy to center, it only cuts a specific number of slices, is difficult to clean, is difficult to store and a cutting board is still needed under the pizza.

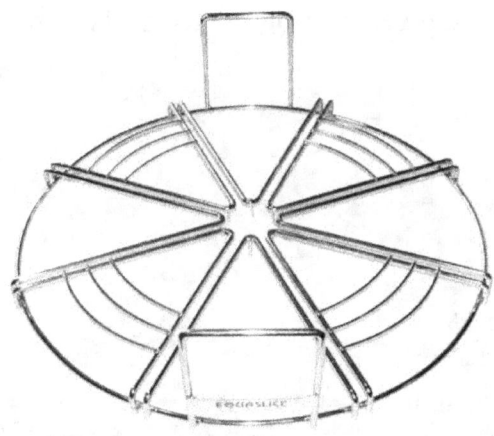

Wire Pizza Cutter Guide

Greg's research with the multi-blade rocker device, shown below, exposed numerous issues which made the cutting device not as user friendly as Greg would want for his employees. For example, it only cuts a specific number of slices which means that several devices would need to be purchased for different pizza sizes and applications. This would present an expensive investment for the buyer as well as having to store all the devices. All of the intricate angles in the blades and frame would make it difficult to clean. Plus the buyer would still need to buy a cutting board for under the pizza cutting device and the pizza.

Multi-Blade Rocker Cutter

The Pizza Slice Cutting guide shown below also poses several issues for the end user. Although it has lines, it is difficult to operate because the pizza covers up a large portion of the lines (not grooves); there are centering dashes but it takes longer to center the pizza based on the dashes, which is especially important during the busy, high volume lunch and dinner hours; and there is no handle to lift the cutting board to slide the pizza off.

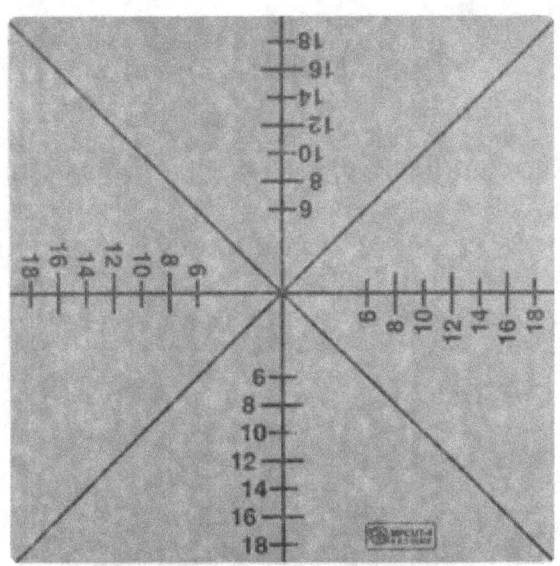

Pizza Slice Cutting Square

The metal Pizza Slice Cutting Device shown below is difficult to operate, difficult to clean and it only cuts a specific number of slices.

Metal Pizza Slice Cutter

The Metal Pizza Slice Cutting Device below has grooves, however, it is difficult to operate because it lacks centering circles, it only cuts a specific number of slices, there are grooves on only one side of the device and there is no handle to lift the cutting device to slide the pizza off.

Metal Pan Pizza Cutter

The Metal Flat Plate Cutting Device below is difficult to operate because it lacks centering circles, it only cuts a specific number of slices, there are grooves on only one side of the device and it dulls the cutting tool.

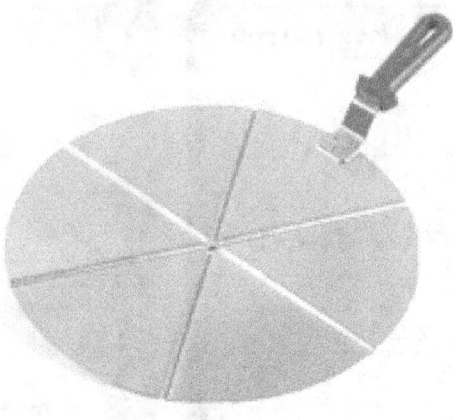

Metal Flat Plate Cutter

Portion PadL's Competition	Your Product's Competition
Wire Pizza Cutter Guide	
Multi-Blade Rocker Cutter	
Pizza Slice Cutting Square	
Metal Pizza Slice Cutter	
Metal Pan Pizza Cutter	
Metal Flat Plate Cutter	

Fill in the notebook, it will help with completing a Detailed Business Plan, by completing the notebooks in the various Volumes of the Portion PadL™

All of the research told Greg that if he wanted a better cutting device than the products that were in the marketplace, he needed to develop his own invention.

He wanted something that would be easy to use and a product that even a new employee could use to cut equal pizza slices the very first time.

He had an idea that he wanted to share with his wife, Alisa, so he drew a crude picture of what he envisioned and said, "This is what I need to solve the problem," and Alisa agreed.

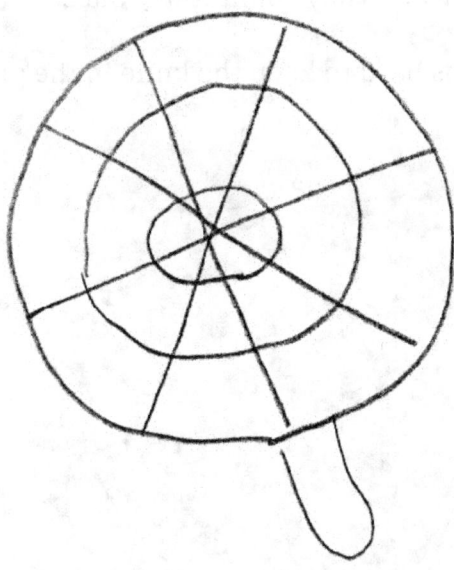

Sketch of the Portion PadL^{tm}

Greg has many skills but drawing and woodworking are not among them so he approached his brother-in-law whose father is an amateur woodworker. Greg showed him the sketch of what he envisioned.

Using Greg's sketch, his brother-in-law's father made the first cutting board out of plywood with rectangular grooves and the circles burned on the board with an electric heating iron. But when Greg tried it out, he found that the rectangular grooves didn't work very well. They splintered, collected food particles, they were difficult to clean, and it was hard to keep the knife in the grooves.

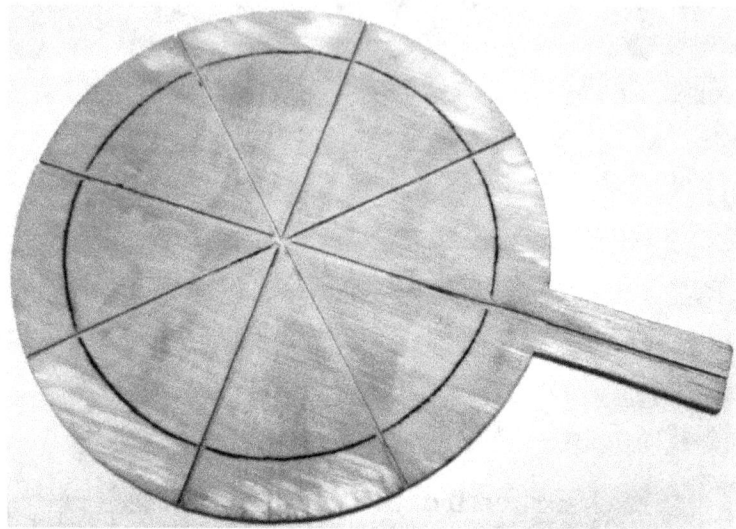

An early wooden version of Greg's pizza cutting board with 4 rectangular slots and a circle burned on one side of the board

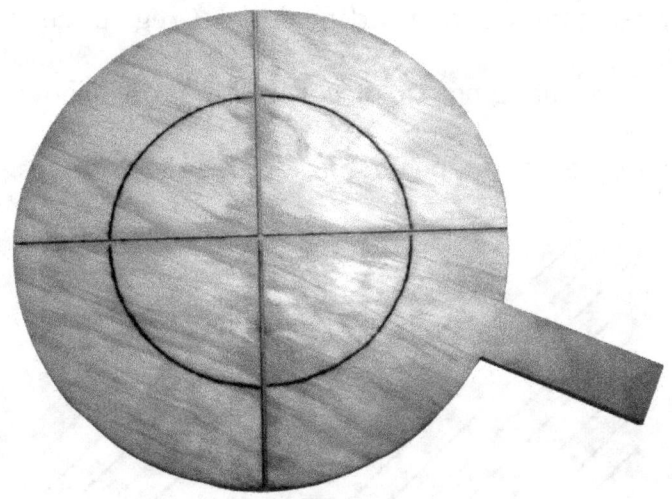

An early wooden version of Greg's pizza cutting board with 2 rectangular

slots and a circle burned on a second side of the board

With the early boards, the slots were cut into the wood with a router.

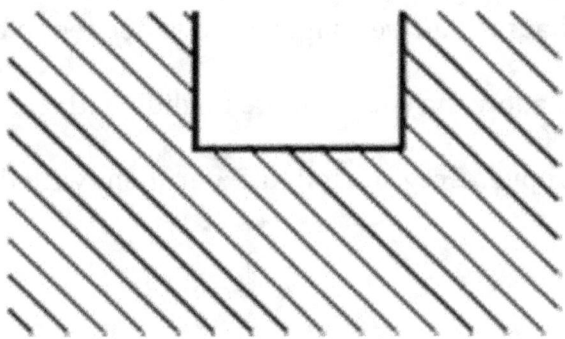

Shown Above: Rectangular Groove

Though the rectangular grooves seemed to be a good idea, his employees were having difficulty getting the cutting knife through the grooves with ease. This problem frustrated his employees because the design slowed them down from cutting the pizzas and keeping up with the fast pace of pizzas coming out of the oven.

After months of using the wooden cutting board, Greg noticed that the rectangular grooves were changing. He discovered that the grooves were getting modified into a V shape from the repeated use of the rocker knives.

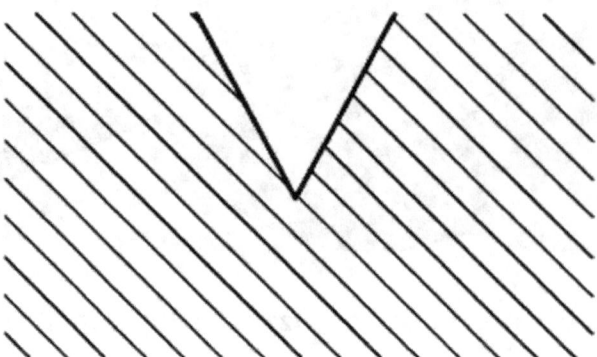

Shown Above: V Groove

He determined that the square shaped groove needed to be a V groove. The V shaped groove would allow the cutting knife to glide with ease and keep the cutting knife in the groove. Greg started to experiment with different shaped and sized V grooves.

With observation and continuous experimentation, Greg made changes to improve his cutting board. He was getting positive feedback from the employees with each improvement.

Greg also noticed that the marked circles that were being used to center his 9, 12 and 14 inch pizzas were bigger than the pizzas that were coming out of the pizza oven.

Greg started to experiment with the centering circles on the pizza cutting board. He measured the size of the centering circles. The circles measured exactly 9, 12 and 14 inches, the advertised size of the pans.

Greg then measured the pizzas coming out of the oven. The pizzas measured approximately half an inch smaller than the circles on the pizza board.   This discovery meant that pizza sizes differ from the advertized pan size.

Notice the circle around the pizza is larger than the pizza

Greg researched the pizza industry, both commercial and residential and discovered many pizzas advertised as the same size vary in actual size (diameter)."   Greg concluded that pizza sizes will vary based upon each customer's pizza diameter, in other words there is no consistent pizza size/diameter in the pizza industry. The pizza cutting boards would need to be custom made to the customer's pizza size applications.

Greg's next series of pizza cutting boards were modified based on his observations and discoveries. He applied the preferred V grooves and reduced the diameter of the centering circles on the board by half an inch.

These further improvements increased his employee's ability to cut the pizzas quicker and resulted in equal size pizza slices.

Greg experimented with plastic, which was approved food safe, but encountered other problems with the material. The plastics proved to be too smooth and the pizzas slid like a hockey puck on ice when cutting the pizzas on the plastic board. In addition, the material's handle wasn't strong enough to support the weight of the pizza.

HDPE (High-Density Polyethylene) Plastic Pizza Cutting Board

Shown below is one of the early boards that were used at Greg's Pizza Business. It was made using laminated food-safe birch. Both the grooves and circles were routed into the pizza board.

Pizza Cutting Board used in Greg's Pizza Business

Pizza Cutting Board with Worn Grooves

Greg also researched combinations of wood, composites, cork and plastic, all common materials that cutting boards are made from.

Birch Laminated Pizza Cutting Board

Even with a relatively simple invention like the pizza cutting board, the research and development stage of an invention is very important. Dozens of prototypes were tested, developed and experimented with before deciding on the correct material and dimensions.

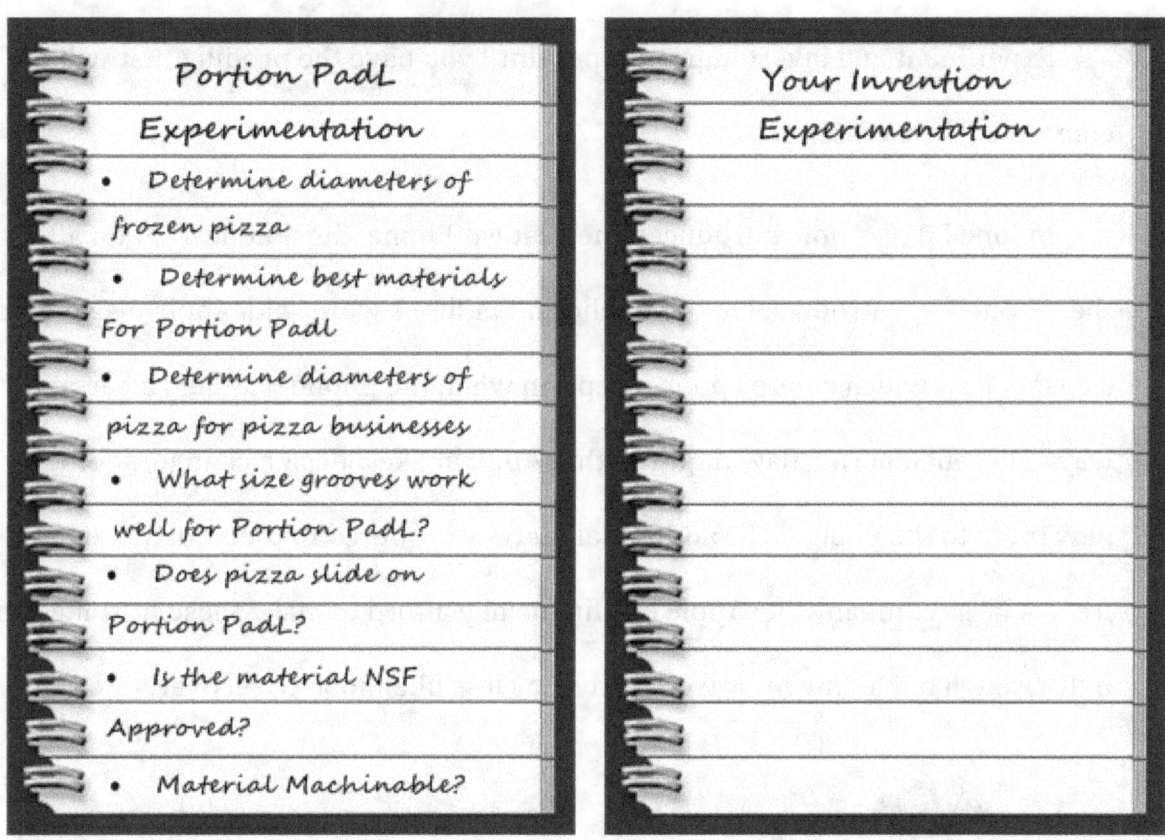

Portion PadL Experimentation	Your Invention Experimentation
• Determine diameters of frozen pizza	
• Determine best materials for Portion Padl	
• Determine diameters of pizza for pizza businesses	
• What size grooves work well for Portion PadL?	
• Does pizza slide on Portion PadL?	
• Is the material NSF Approved?	
• Material Machinable?	

Fill in these notebooks it will help with completing a Business Plan when you complete the various Volumes of this Series on the Portion PadL™

ANOTHER VOLUME: We will describe in greater detail in another volume how to work with prototype companies and manufacturers of the product.

Experiment and invest in prototypes until you have the product that will be commercially successful.

In June 2010, Apple introduced their latest iPhone, the iPhone 4. Soon after the iPhone was introduced to the public and millions were sold, Apple discovered a design flaw which created poor reception when the phone was held a certain way. This engineering flaw required that Apple make a design change, adding a new piece to the phone. The additional piece was offered to their customers for free. A large company like Apple can financially afford to make these mistakes and pull through but an inventor would have far less likelihood to recover.

Spilled Coffee Chronicles Advisor Sez™:

Invest the time and money to get the product right the first time!

Even with a simple invention like the pizza cutting board, the research and development stage of an invention is very important, time consuming and expensive.   Dozens of prototypes were used and tested by Greg before he decided on the correct material and was convinced he had the right product to release. (Note: In the end, the market will tell you whether you have the right product because people will either buy it or not). As stated before, it is critical to start with an ingenious/elegant yet simple product because even a simple idea turned into a commercialized product requires a lot of time and money.

Greg began using the cutting board in his business. It was easy to use and did the job; and his institutional customers loved the perfectly sliced pizzas. They no longer had to throw away smaller slices that no one wanted and students moved through the cafeteria lines faster because they didn't waste time looking for the biggest piece.

Spilled Coffee Chronicles Advisor Sez^{tm}: Get to positive cash flow as soon as possible

You want to get to positive cash flow as soon as possible. Complex inventions are difficult for large companies to develop even with large financial resources and various departments including marketing, engineering, sales, prototype shops and consultants. One of the biggest reasons inventors fail is they run out of money developing complex inventions.

Entrepreneur Stories: Dyson Vacuum

As mentioned before, prototyping a product is an expensive, long and painstaking process.

 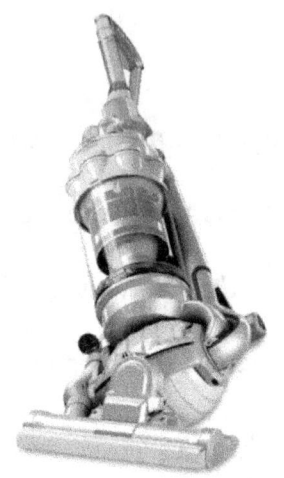

After **15 years** of tinkering and **5,000 prototypes**, James Dyson (a brilliant and visionary inventor) developed a complicated bag-less, cyclone vacuum — the first vacuum that doesn't lose suction. Very few inventors have the persistence and are able to raise the funds like Dyson.

The Dyson vacuum cleaner is a great example of a product that took a long time to achieve success. According to www.dyson.com, James Dyson began to develop the first cyclonic bag-less vacuum in 1978. He offered his invention to all the major vacuum manufacturers. They turned him down. Why? The vacuum manufacturers made hundreds of millions of dollars every year selling replacement bags. But James Dyson didn't give up. After several years his idea eventually caught on and his company has developed products that have achieved sales of over $10 billion worldwide.

www.dyson.com

The improved pizza cutting board made it easy to calculate a fundraiser's profit. Since there was no waste, they could accurately calculate how much a fundraiser would make on each pizza.

Pizza Cutting Board
Volume 1, Chapter 5
What to do with your invention?

It's not enough that someone says they like the invention or product; they have to be willing to pay for it. A lot of inventors obtain a patent for their products, but they don't know how to market it or if there is even a market for the product.

It's like American Idol where all the contestants think they are the best singer in the nation. Then they open their mouths and it sounds like nails scratching on a chalkboard. They don't see their own shortcomings. Inventors often can be the same way.

Ask an inventor if he's smarter than other people and 90 percent will say they are. They often think because they invented the product, they know more about it, and they are the best person to market and present the product to others. But more often than not, they are the last person that should try to sell their own invention.

They just don't know how to do it.

It really crushes any chance of success in the long term because they don't know how to commercialize something. They are too close to the invention to see it in an objective manner. They can't see the product through the eyes of the consumer, the investors or the key decision makers.

Often the first things an investor group will do if they agree to take on an inventor's product is to replace the inventor with a chief executive officer (CEO). It's not that the inventor isn't important, but after the product is made, they want someone with experience in selling, not making it.

Inventors need to know where the gaps are in their abilities and how to fill those gaps. Greg had to take a hard look at what skills he possessed and brought to not only inventing a product but producing and marketing the product.

Pizza Cutting Board
Volume 1, Chapter 6
Finding the Right Business Partners

The grooved pizza cutting board helped Greg keep his customers happy which led to increased sales for his pizza business. His problem for cutting equal size slices was solved for the most part. But it didn't take long for Greg to realize that improving his own business was only part of the picture. He already knew from his research that every other pizza shop had the *same* problem. Could there be a market for his invention? What improvements had to be made to the existing board? How would he have the board manufactured? How would he sell it? Would he need a patent?

Greg didn't know the answers to these questions, so he did more research. He knew that patents were valuable and found that a patent is a grant made by a government that gave the sole right to make, use, and sell that invention for a set period of time. He decided that the next step would be to get a patent.

Greg and Alisa contacted a patent attorney in downtown Cleveland, Ohio, and made an appointment to see him. The attorney told them that doing a patent search and applying for a patent would cost thousands of dollars. But since the attorney had no idea how to manufacture and sell a product, the meeting was disappointing and discouraging.

That night Greg and Alisa talked it over. "We already have our hands full with the pizza business," Alisa said. "We can't start something else right now."

"But I think there could really be a market for it," Greg said. "There's nothing else like it."

"I know, but it'll cost thousands to get a patent and that's just one of the first steps. Then we'll have to find someone to make the cutting boards, advertise them and distribute them. The investment could be huge and it might be years before we show a profit."

Finally, they decided to put the idea on hold. Greg would continue to use various versions of the cutting boards in his pizza business, making improvements, but getting a patent and selling the cutting boards would have to wait.

Greg's dream of commercializing his cutting board languished as he focused on his pizza business. Then something happened which again got him thinking about commercializing his pizza cutting board.

Greg belongs to several networking groups composed of professional men and women who meet socially to exchange ideas and help each other in their careers. In October 2009, one of these groups held a chili cook-off. Participants were asked to bring some of their favorite chili and at the end of the event everyone would cast their vote for the best recipe. Greg attended the chili cook-off, but instead of chili, he brought a chili-pizza that was served on his pizza cutting board.

At the event, Greg met another patent attorney, Andrew Spriegel, who had his law firm in the same building in Hudson, Ohio, where the event took place. When he found out that Andrew was a patent attorney, the conversation turned to patents. Andrew thought Greg's pizza board was a great product and could be commercialized. As the two men got to know each other, Andrew saw Greg's marketing ability and Greg recognized that Andrew had the foresight and connections to help him commercialize the cutting board.

Finding the right Business Partner	Your Answers Regarding a Partner
• Do you need a business Partner?	
• Have you checked his/her Background/References	
• Is he/she driven to succeed?	
• Is he/she willing to put all agreements in writing?	
• Are you aware that a partnership is like a marriage?	
• Is everyone aware of their contributions?	

Fill in the notebook

The pizza cutting board met all of Andrew's criteria to be a successful invention:

Criteria	The Equal Size Pizza Cutting Board
It could be patented	Having a patent is often the key to the successful commercializing of a product. If the product is successful you do not want others to copy it.
It was simple	Inventing is difficult enough even with a simple product.
The development could be self funded and did not require investors (angel investors or venture capital)	If your product requires funding then it can come to a halt until you find an investor. With self funding, the project cannot be stopped unless the inventor & partners give up trying to find buyers to buy the product.
It was breakthrough technology	There was nothing in the pizza marketplace that worked as well as the Portion PadL[tm].
It didn't require tooling.	Tooling is very expensive and often requires a large investment and large inventories of product.
Greg was hard working and already a successful businessman in a demanding business (the pizza and food business)	Inventing is a difficult undertaking and requires a lot of hard work. Greg had a strong work ethic, was realistic and knew the work a successful business required.
The product could be sold before it was manufactured. In other words, an order could be taken and be paid for upfront and then sent to the manufacturer to have the board made.	This gets positive cash flow very quickly. However, the manufacturer has to be willing to make small quantities in the beginning until the number of orders pick up. (Not easy to do)

Andrew knew the people that would manufacture the product and laser etch the logo.	It is good to know manufacturers that will work with them. It is not absolutely necessary, but it helps.
Greg had his wife Alisa's support in commercializing the product.	This goes a long way toward the success of the project if your family is behind your efforts.

How about your invention or product?

Criteria	The Equal Size Pizza Cutting Board
Can it be patented?	
Is your invention simple?	
Can your invention be self funded or does it require investors?	
Is your invention breakthrough technology?	
Does your product require tooling?	
Are you hard working and do you finish tasks once you start them?	

Can your product be sold before it is manufactured?	
Do you know manufacturers?	
Do you have the support of your spouse and/or family?"	

Fill in this Chart

Greg and Alisa met with Andrew to discuss working together on commercializing the pizza cutting board where Andrew shared his inventing and commercializing philosophy with them.

Spilled Coffee Chronicles Advisor Sez:

Make sure your spouse/family is involved with the decision to commercialize a product or invention.

It takes a lot of time, energy and money to commercialize a product and your spouse and/or family needs to be a part of the process. When you talk with a patent attorney or advisor, ask them about their success rate with inventors. Do they have a network of consultants that can help you with your invention?

Andrew showed both Greg and Alisa products that he was commercializing. Greg was smart to bring his spouse to help him decide whether he wanted to move forward. It is critical that spouses agree on what they want to do and whether they want to move forward together.

Often before they start the process, inventors are doomed to failure because the product they are trying to commercialize is too complicated. It is difficult for a company to take a complex product to market and nearly impossible for an individual inventor to do so. Inventors need to come up with simple products that will make money, build on their successes and develop more complicated products as cash flow becomes manageable.

Failure is not an Option

When Andrew meets with a client (that is a first time inventor) who wants a patent on an invention, he evaluates the complexity of the idea. He looks at their invention, and if it's complicated, he usually discourages them from trying to get a patent. Sometimes people are shocked when he tells them not to patent their product.

"It is hard enough for a company, with all its resources, money and manpower to get an idea launched, let alone a single inventor. The patent, although expensive, is just a small cost of commercializing the invention. Inventors can invest thousands of dollars on a product and never see a return."

Too many inventors try to create a product that requires investment in tooling, which has expensive set-up costs and requires a large minimal order of the manufactured product from the beginning. Often inventors are stuck with expensive equipment and tools and a large inventory of product they cannot sell.

Injection Molding Tooling is Very Expensive

Andrew suggests that **instead** of starting out with a high amount of product that has a low price per unit (this is a typical approach), keep inventory low, avoid tooling and allow the price of the product to be higher to begin with.

You will initially make less profit per part, but you will get to positive cash flow much more quickly. In addition, if the product doesn't sell well, you haven't spent a lot of money, and you can move on to another product. You don't want to run out of money spending it on inventory before you verify sales. Increase sales first. You will make less profit per part but you will get to positive cash flow more quickly.

Inventors have invested tens of thousands, even hundreds of thousands of dollars on an invention, but have not sold a single product.

Every invention has expenses such as patent costs, manufacturing, material costs, packaging, shipping, a website, insurance, company formation and more. The costs add up quickly. Each expense may not be much by itself, but when added together they can cost thousands of dollars, and the inventor will often run out of money. Keep the upfront investment as small as possible. As mentioned previously, although it costs more to produce small quantities, and there is less profit, there is also less debt and again you reach positive cash flow more quickly.

Andrew's advice was just as valuable as the patent Greg sought to obtain. He brought to the partnership his experience at solving problems. Even small problems can cost big bucks and need a solution, especially for products that will be used by the public.

Andrew also explained his philosophy that it is better to 'fail fast and fail early.' In other words, find out as soon as possible if the product will be a success or failure.

If an invention fails right away without much money invested, it lessens the financial burden. The inventor can attempt to commercialize another product. Some inventors invest a lot of money, and when the invention fails, they never try again. Greg and Andrew agreed to be equal partners in the new company, NuVo Grand, LLC. (Grand = Greg and Andrew).

Inventors also need to be careful not to become so attached to an invention that they can't abandon ship when the idea can no longer stay afloat.

Pizza Cutting Board
Volume 1, Chapter 7
Business Plan

A business plan is a road map for your business and its future. Writing a business plan is often overwhelming to an individual starting a new business. **The various volumes of the Spilled Coffee Chronicles will help you develop a detailed plan and make the process much easier.**

DEFINITION – *According to www.entreprenuer.com a business plan is defined as a written document describing the nature of the business, the sales and marketing strategy, and the financial background, and containing a projected profit and loss statement.*

Business plans are designed to perform a number of tasks for both those who write them and those who read them. They can be used by entrepreneurs seeking outside investments to communicate to potential financial backers. They can also be used to attract valuable employees and help attract new business opportunities. They help in dealing with banks, buyers and suppliers and in planning out the business approach to success. It is best to think of a business plan as a living document, changing and evolving on a frequent basis.

Many people think a business plan is something meant only for a big business, but just as importantly, it is needed for a small business. A business plan is meant to target the strengths of your business to potential opportunities. To do this effectively, you need to research, collect, screen, and analyze information about the business environment. You also need to have a clear understanding of your business and your strengths and weaknesses and develop a clear mission, goals, and objectives and to change those as the business changes. Acquiring this is difficult and requires more work than expected. The Spilled Coffee Chronicles has various work sheets and check lists to help you create a business plan with greater ease. You must realistically assess the business that you want to offer.

As technology and competition evolve and the business environment becomes less stable and less predictable, it is critical that you keep your business focused on the right track. If you are to endure and flourish, you should take the time to identify the position in which you are most likely to be successful and to identify the resources that must be met.

Here is a list of subjects that are normally covered in a business plan:

1. **<u>Table of Contents</u>**

2. **<u>Executive Summary</u>** - The executive summary section of a business plan is a brief summary of the highlights of your business plan. Even though the summary appears first in the business plan, most plan developers leave the writing of the executive summary until the end of the process. This summary is an overview of the rest of the plan. If

you don't get it right, your target audience will not go further than the executive summary. It reports in such a way that business people, financial investors, *etc* can become acquainted with a large amount of material without having to read the entire business plan. It usually contains background information, concise analysis and main conclusions. It is intended to aid decision makers, business and financial professionals regarding their decisions about the business.

3. **General Company Description** — The company description section of the business plan is typically the first section of the plan that is written. This section follows the executive summary. A well drafted company description will help show a general direction of the business and help any potential lenders or partners decide whether they want to be involved in your business, products and services.

4. **Marketing Plan** — A marketing plan is a document that explains the necessary approaches to achieve your businesses marketing objectives. It can involve a product or service, a brand, a product line, a roll out strategy, *etc*. Marketing plans can cover months or years of activity. The marketing plan should be part of the overall business plan. A solid marketing strategy is the foundation of a well-written business plan.

5. **Operational Plan** — An operational plan explains the daily operation of the business. It defines its location, the company's equipment and value, an organization chart, the various processes needed to run the business on a day to day basis, and other operations.

6. **Management and Organization** — The M & O establishes written standards to measure individual performance. The company needs to define goals for organizational departments in specific, operational terms that include standards of performance to compare with organizational activities. In addition, the actual performance needs to

be measured using measurement forms. For example, if sales growth is a goal, the organization should have a way of gathering and reviewing sales data.

7. **Personal Financial Statement** — The objective of a personal financial statement is to provide information about your personal financial position. Financial statements should be understandable, relevant, reliable and comparable. Reported assets, liabilities, equity, income and expenses are directly related to your personal financial position and not the financial position of the company.

8. **Startup Expenses and Capitalization** — Starting a business always requires capital to cover expenses to get a business of the ground. A company needs to carefully estimate expenses and then to document where you will get sufficient capital to finance the start-up. If your research is comprehensive, there is less chance that you will accidentally leave out important expenses and/or underestimate them.

9. **Financial Plan** — a financial plan often refers to the three primary financial statements, a balance sheet, an income statement, and cash flow statements.

OTHER VOLUMES: Business Plan items 1-9 above will be explained in detail along with fill-in charts in other volumes.

Pizza Cutting Board
Volume 1, Chapter 8: Was it Attempted Theft of an Invention? You make the Call!

During this time, Greg did more research on the best material to use for the boards. Although the wood boards were working well in his own pizza business, he realized that cutting boards sold to commercial establishments had to be more durable. He also knew that cutting boards had to meet certain health standards (NSF) for the boards to be widely accepted. For example, the board used in his shop and the ones he was experimenting with had to be food-safe. The commercial boards had to be water resistant, temperature resistant, bacteria resistant, dishwasher safe and meet NSF food equipment specifications.

Greg told Andrew they needed to find an NSF approved material and the wood would not be a suitable commercial material. NSF was founded as the National Sanitation Foundation in 1944 to standardize sanitation and food safety. (Note: at this point Greg and Andrew started working together forming a limited liability company NuVo Grand, LLC where they were equal members.) NSF International assists companies in meeting federal requirements and improving their food safety systems beyond federal regulations. Andrew said that wood is used all the time to make cutting boards. Besides, finding another material would set back the timeline for introducing the cutting board to the public. However, Andrew listened because Greg knew the food business. The change was important as stated before because large pizza franchises and food companies will only buy commercially food safe products (NSF approved), that are easy to use and easy to clean.

Greg went online and found a composite of wood fibers and poly resin, which was developed for skateboard ramps. It is a hard material, resists knife scratches, has a high tolerance for heat and is dishwasher safe. It was an ideal product for the Portion PadL^{tm} cutting board that met all of the identified qualifications. Greg contacted a large cutting board manufacturer, one of the few companies that make boards using the composite.

At this point Greg did what Andrew had warned him against on several occasions, giving away too much information. In Andrew's experience as a patent attorney and as an inventor he has seen companies steal or try to steal someone else's idea or invention. As an inventor you need to be very careful about those with whom you share information. It is critical to sign non-disclosure agreements and keep information to yourself. You should not talk with other companies or individuals unless they have signed a non-disclosure agreement.

ANOTHER VOLUME: Non-disclosure Agreements will be covered in detail in another volume. There are a few critical paragraphs that are critical to Non-disclosure Agreements.

The composite could only be purchased through authorized dealers in 4-foot-by-8-foot sheets. Greg needed to purchase some of the composite to determine if a cutting board could be made from it and if it could be manufactured." Greg, also, needed to test the board. "I wanted to make sure I had the right material," Greg said. "I didn't know if any material was durable enough and food safe to work with until I tried it." He bought a rectangular composite cutting board already manufactured from an authorized dealer to alter and create his own board and test it. After testing the prototype board in his store, Greg determined it was the right material for the application. It worked.

Unfortunately inventors are proud of their invention and like to talk about them, but until a patent is in place to protect an invention, ideas can be stolen. Luckily, Andrew had written and filed two utility patents when Greg learned the lesson the hard way about sharing information and trusting business people.

"When I ordered the composite material, I talked to the CEO of a large pizza board manufacturing company and explained what I was using it for," Greg said. "He seemed very interested in it."

The CEO suggested the company make the cutting board for Greg and pay him a royalty for everything they sold. "It sounded interesting, and I explained" Greg said.

They had several more telephone conversations about his invention but after the fifth phone conversation, the CEO told him there may be a "conflict of interest" because they were going to develop a cutting board very similar to Greg's board.

The CEO e-mailed drawings of the cutting board to Greg that came out of their conversations. The drawings included all of the features that Greg had described about his cutting board. As we like to say, the "ink was still wet on the drawings."

"It was very shocking to see how a company was so quick to copy a product," Greg said. "I learned I gave too much information away to a competitor."

Greg realized an inventor can't give away a lot of information until a patent is pending. Even with a patent pending, you need to be careful about sharing information with others." "I almost screwed up by talking too much too early," Greg admitted. "Every part of the inventor's journey is a learning venture — learn as you go."

The drawings were similar to his original cutting board but not the current cutting board. Luckily, Greg didn't give away all the specifics of the Portion PadL™ to the CEO or tell him about the changes, improvements and patentable features he had added during experimentation.

At this point Greg and Andrew had two patents pending on the Portion PadL™ and so the CEO cut off discussions with NuVo Grand, LLC. Greg had made a mistake, but Andrew suggested he avoid dwelling on it. They needed composite material, but he didn't want to buy from the CEO. He called the manufacturer of composite material, and they gave him the location of the local distributor in Macedonia, Ohio, which was nearby.

"The local distributor worked out much better," Greg said.

Pizza Cutting Board
Volume 1, Chapter 9: Unforeseen Obstacles

There will be many obstacles on the path to success. There will be emotionally high and low days; the key is to keep moving forward.

The Snowstorm

Greg could purchase the 4-by-8 sheets of composite material in Macedonia and take them to Fredericksburg, where the boards were being made. It was just a 40-minute drive.

Greg found someone in Barberton who could do the machining needed for making the cutting grooves and measuring circles. He could also package and ship the finished product from Barberton. Now they had two manufacturers on board.

Think Outside the Box

When you are faced with **challenges** you often have to **think outside the box**.

You will face obstacles and issues along the way and you have to keep moving

along through the muck and the constant ups and downs.

It took Greg several months of using various cutting boards made of different materials at his pizza business before he decided on the right material for his cutting board. Finally, Greg was satisfied. An unexpected benefit of using the composite material was that the CNC manufacturing company said the material "cut like butter" when making grooves, as opposed to wood.

Ten Equally Cut Pizza Slices using the Patent Pending Portion PadL™

US Patent

In the meantime, Andrew knew several companies that could make the cutting boards. Ohio has a large population of Amish who are known for their craft skills. Andrew had connections in the Amish community because he had helped an Amish man with his invention and had been introduced to Amish manufacturers. He made arrangements to have the composite panels cut into Portion PadL™ cutting boards.

Amish Buggy

But someone had to get the sheets of composite board to the Amish, who do not use cars or trucks. The trip from the supplier to the Amish craftsmen was only about 40 miles. So, one day in early January of 2010, Greg ordered two sheets of composite board and decided to transport them, strapping them to the top of his SUV.

Unfortunately, complications developed. Greg didn't know the sheets of composite board were so heavy. After only a few miles, he realized that his SUV couldn't handle the weight of the boards on top of it. Also, the wind was picking up and it began snowing, making driving the SUV too dangerous to continue. Greg was facing one of numerous obstacles that Andrew and Greg have faced along the way.

Greg decided to stop at a nearby U-Haul truck rental company and continued the trip with the composite board sheets in a rented panel truck. But his ordeal was just beginning.

What Greg didn't know was that the biggest snowstorm of the season was bearing down on Northeast Ohio. Visibility approached near zero as windblown snow accumulated at an alarming rate, but Greg pushed on. He was determined to get the composite board panels to the Amish craftsmen and return home that same day.

The severe weather conditions forced Greg to slow the truck to a crawl. By the time he reached Amish country, night had fallen.

Amish country is mostly rural — there are few towns, and they are separated by many miles of unlit, unplowed country roads. Adding to his difficulty was the fact that few of the roads have street signs or street lighting. Greg soon realized he was lost.

At one point he got stuck when he tried to turn the truck around. Luckily, an Amish family was passing by, and they helped him get going and gave him directions.

Eventually, Greg found the Amish craftsman's business and dropped off the composite boards so they could be made into cutting boards. He headed back to the truck rental company to retrieve his SUV.

But the storm did not let up. By the time he returned the truck and made it safely home at 11 p.m., over ten inches of snow had fallen. A trip he calculated would take only three hours, actually took thirteen hours.

Pizza Cutting Board

Volume 1, Chapter 10: The proper name/logo selection is critical.

Why are names, logos, and trademarks so valuable?

A trademark is one of the most valuable assets that a business can create and yet very few businesses take advantage of this intangible asset. Remember, you want to distinguish yourself from your competitors, and every time you create an advertisement or perform excellent service, you are building value into your business name and/or trademark.

When starting a business there are very few tasks more important than selecting an appropriate company or product name. Names can be distinctive (Xerox), unique (Google) or descriptive (Joe's Barber Shop). Try brainstorming names. Take the time to make a list of words.

Distinctive names receive more trademark protection. Distinctive business names (such as Xerox, Ditch Witch, 1-800-Flowers, Quicken Loans, and

Amazon.com) are clever and memorable, and they usually receive a trademark under federal and state trademark law.

When XEROX first came out no one knew how to pronounce the Name!

Great Name: What do ditches have to do with witches? Nothing but the name is novel, catchy and it says "our equipment digs ditches." It is also easy to remember.

1-800-flowers.com™ is one of the best known Trademarks in the US

Common or ordinary names (such as Smith's Hardware, Tom's Gourmet Sandwiches, and Joe.com) usually do not make good Trademark names.

Why is a distinctive name important?

1. Your name immediately comes up high in the search engines.

   a. We could have named the pizza cutting board the Portion Paddle rather than our name, the Portion PadL.   However, if you search Portion Paddle in Google, you will get results like canoes, kayaks, spanking, ping pong, *etc.*

   b. When you type "Portion PadL" the results on the first two and one half pages on Google are topics related to our product or business.

2. It is much easier to distinguish your products and services from other products and services.

   a. Which product name of the following pair of product names do you think distinguishes itself from the other product name; Under Armour or Joe's Tee Shirts; Snap-on tools or Heavy Duty Tools and Brothers Diner or Mickey's Dining Car.

3. A Trademark is a measure of a company's good will and can become your most valuable asset!

a. Andrew recently had a client that does auto wraps, the advertising vinyl coating that is applied to the vehicle. The name they used is Auto Pro Design Graphics. Not only is it generic but is difficult to remember and it would have very little value when the owner retires.

b. Andrew convinced the owner to change the name, and they came up with the "Wrap Professors". Because of the uniqueness of the name and the fact that it is easy to remember, the owner can eventually franchise the business or sell the name along with the business when he retires. He can also incorporate the previous name with the new name using a tag line.

c. "The Wrap Professors, we put the **Pro** in Auto Pro Design." Using this strategy, once customers associate the two names as being one business he can drop the name "Auto Pro Design."

Here was the process we used to name the pizza cutting board, the **"Portion PadL."** We made up lists of names to describe an equal slice pizza cutting board. Here are the parameters we used to come up with a name:

1. Did it conjure up the vision of an equal slice pizza cutting board?

2. Was it fanciful and/or unique enough to come up high in the search engines?

3. Did the name work well with a logo?

4. Was the name impressive?

5. Was it easy to remember?

6. Was it easy to spell?

We tried hundreds of names. Here is a list of some of the names that we came up with:

a. The PreciseSlice pizza (Negative: If you type "PreciseSlice pizza" into Google too many results pop up)

b. PizzaPrecise cutting board (Negative: The name was too generic to rate high in the search engines)

c. PS101 (Negative: It didn't conjure up the vision of an equal slice pizza cutting board, and it sounded more like a public school)

d. Pizzaprezize (Negative: It wasn't easy to spell or remember)

e. SureSlice (Positive: It had potential)

f. Diamond slice (Negative: Too confusing and it didn't remind someone of pizza slices)

g. Pizza Trax cutting board (Positive: We liked this name with the play on tracks and we thought it would work well with a logo. Negative: It didn't remind someone of pizza slices)

h. The Pizza PaL (Positive: It had potential)

i. The Pizza Precizer (Negative: It wasn't easy to spell or remember)

j. Pizzaccurate

k. Kleenkut

l. Da right slice

m. Da riteslice

n. Slice trax

o. EZ Trax

p. Railrider perfect slice cutting board

q. Pizzarite

r. Equaslicer

s. The pizza diva

t. Exact slice

u. Exact cut

v. Deco board

w. The grate divide

x. Ez Slice

y. Ez Cut

z. The Trucut

aa. The Ez Action

bb. The EzGlide

cc. EquaSlice

dd. TruKut

ee. CleanKut

ff. NoBrainR

gg. No BrainR

hh. Confidence

ii. Utopia

jj. Pizza Pizzaz

kk. Pizzaz

ll. Groovy

mm. GrooV

nn. GrooVee

oo. Groove V

pp. Pequal

qq. GruV cut

rr. The equalier

This will probably appear to be way too names to most people, however large companies often spend months or years picking the right name.

Again, naming the product is one of the most important choices an inventor can make. It is like naming a baby. The name stays with that person (product) for the rest of their life. "Then it came to me, and it sounded cool — GrooVkut," Greg said.

Although Greg liked it, Andrew did not because the name didn't identify what the product did, and he said it sounded like "bell-bottoms and beads". Andrew came up with the name — Portion PadL. The product sounds like a paddle, but is novel, would come up high on the search engines and the board cuts equal portions so the name clearly described the product.

Luckily, Greg liked the name as well.

"We both agreed on the Portion PadL. Once we came up with that, we worked on the logo. "

They were eager to brand their product and make it their own. Here are some of the logos that they were experimented with:

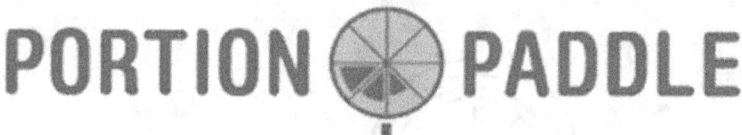

The lettering above was submitted by a marketing consultant, it was a great start, but they struggled with several features of the logo. It was decided that the lettering looked too old fashioned so the lettering was changed to a more contemporary style. It was also missing the circles which are a key feature of the PadL.

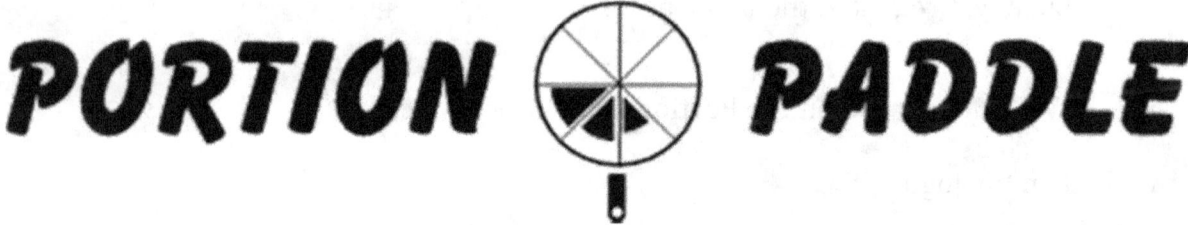

In addition, Andrew and Greg decided to enlarge the logo and add the centering circles, against the consultant's advice. In the end they were much happier with the end result, below. Another big feature was to change the name from Paddle to "PadL".

EVERY SLICE IS A PROFIT CENTER

Andrew knows it is critical in creating a trademark to make the name fanciful ("made up"). The trademarks that often have the highest estimated worth over time are fanciful. Therefore, it was critical to change the word "Paddle" to "PadL". One of the big advantages of using a fanciful name is that you immediately come up higher on the search engines. If you type "Portion Paddle" into Google, up comes canoes, kayaks, spanking paddles, ping pong paddles, *etc.* If you type in "Portion PadL", up pops our product or our business name. Right away you have reduced the need for search engine optimization (SEO).

Here are a few of the estimated values of several well known trademarks:

$27.9 Billion

$67 Billion

$56.2 Billion

$56.9 Billion

The best trademarks are instantly recognized and conjure up in the minds of potential customers, quality, dependability, *etc.*, of the source of the goods or services. Give a lot of thought to the name of your product or company.

> **DEFINITION** – A trademark is a word, slogan, symbol, design, or combination of these elements, which identifies and distinguishes the goods or services of one party from others.

Here are some suggestions:

1. Create a fanciful name, for example, by putting together two common words (of course choose words that convey the spirit of your business) like BlueDog or BigVoice.

2. Choose a name inspired by words from another language.

3. Think of a name inspired by a geographic location.

4. Ask your friends and family for ideas.

5. Put together descriptive words related to your industry, like Whole Foods or PetWash Station.

6. This is not an easy process and usually takes a lot of time.

Entrepreneur Stories: Nike "Swoosh"

Phil Knight launched a new line of shoes, protected by the new name, Nike. In 1971, he commissioned a student named Caroline Davidson to design a logo for this new product line. She drew up the famous "swoosh," for which she was paid $35.

Pizza Cutting Board
Volume 1, Chapter 11: The Pizza Cutting Video

When Andrew and Greg initially started to sell the Portion PadL™, one of the issues that they had was that everyone wanted a free board so they could test out the cutting board. Andrew insisted that they create a video to show the Portion PadL™ in action. Greg finally gave in and they created a video in Greg's business on a very busy night. Greg staged the pizzas so that they were coming out of the oven even more quickly than normal and they were able to show how effective the Portion PadL™ was in action. After releasing the video, potential customers stopped asking for a free board and businesses started purchasing them. The key was for people to see the Portion PadL^{tm} in action. Up until that point the Portion PadL^{tm} was an abstract thought, but after the video, they could clearly see how effective it was on the job.

Entrepreneur Stories: Home Depot

Sometimes people have to actually see your vision.

Home Depot was founded by Bernie Marcus and Arthur Blank. They opened the first two Home Depot stores in Atlanta, Georgia. The first stores were huge warehouses that dwarfed the competition and stocked more products than the average hardware store at that time. In order to give the illusion of a warehouse full of products, they piled empty boxes, above the customers reach, all the way to the ceiling. Up until that point people could not conceive their vision.

www.homedepot.com

Pizza Cutting Board

Volume 1, Chapter 12: Communication is KEY

Spilled Coffee Chronicles Advisor Sez[tm]:

Whenever you communicate with someone try to remember this well known communication study.

In 1990, Elizabeth Newton, a Stanford University graduate student in psychology, demonstrated the criticality of context in communication by utilizing a simple game in which she assigned pairs of people to one of two roles: a "tapper" or a "listener." Each tapper was instructed to pick a well-known song, such as "Happy Birthday," and tap out the rhythm of the song on a table. The listener was instructed to guess the song.

Over the course of Newton's experiment, 120 songs were tapped. Listeners guessed only three of the songs correctly (or 2.5%). However, before the listeners guessed the song, the tappers were asked to predict the probability that listeners would guess the song correctly. They predicted over 50%. The tappers got their message across one time in 40, but they thought they would get it across 50% of the time. Why?

When a tapper taps out a song, he/she hears the tune playing along to her taps. Meanwhile, all the listener can hear is tapping. Yet the tappers were shocked by how hard the listeners had to work to pick up the tune.

Therefore it is critical when working with customers, suppliers, and manufacturers to communicate in concrete language. That is why it is critical at times in the conversation to ask what the other person is hearing to ensure that they are getting the **CONTEXT** of what you are saying. Miscommunication is often the result of conveying information too quickly and assuming that the other person knows what you mean.

Miscommunication can be very expensive, such as getting the dimensions wrong on a part, missed orders, late deliveries, and similar issues.

Communication

Pizza Cutting Board

Volume 1, Chapter 13: Shipping and Handling the Early Stages.

The product was ready but Greg and Andrew had a problem with shipping. The handle on the cutting board made it difficult to fit into a conventional shipping container or box.

"I tested what kind of boxes would work," Greg said. "Pizza boxes did not work very well because of size."

Greg discovered that the boxes that the pizza cheese came in at his restaurant would work by cutting one end and sandwiching the cutting board between the cardboard.

"We shipped them in the empty cheese boxes," Greg said. He needed someone to be the courier and transport the cutting boards. He ran an ad for a driver to pick up the sheets of composite board and drive them 35 miles to the manufacturer. His general manager at the pizza business, Will Shaw, suggested asking a customer who is also a small business owner. Due to the hard economic times, his business was struggling; he might want to make a couple extra dollars. Greg could have brushed off the suggestion, but instead, he called up the customer, who owned a large white van and was willing to do the work. Greg met him at the warehouse and showed him the route from start to finish. When Greg asked his new courier how much he wanted paid, he said 20 large pizzas and the price was right!

NuVo Grand, LLC had a product, the Portion PadL, ready for production that could be sold on the commercial side of the business. Greg and Andrew started working together in December of 2009. Four months later, the Portion PadL was introduced to the commercial market.

This timeline and success is not commonplace, but it can be accomplished. The Spilled Coffee Chronicles is intended to share with you the proper steps, including the mistakes made, so you can successfully commercialize your idea.

The journey for the pizza cutting board doesn't end here. Each chapter of this book is a small snapshot of the journey.

When you decide to go forward with your idea, you will probably go through similar steps or the stages that Greg and Andrew did. Each volume of the Spilled Coffee Chronicles, The Pizza Cutting board, will address all the stages you'll encounter while pursuing your idea into an invention.

End of Volume 1, additional volumes will be coming out in the near future for the Pizza Cutting Board.

About the Main Author

Laura Freeman grew up in Stow, Ohio, and attended Kent State University for several years before getting married and raising two sons in neighboring Cuyahoga Falls. In her spare time she played women's softball in the as a pitcher and played outfielder for eight years and was on the offensive line of the Kent Twisters, a women's ice hockey team for seven years.

She returned to Kent State University in 2002 and graduated with a bachelor's degree in communications and a minor in creative writing in December 2004. She began working for the Record Publishing Company in February 2005 as a reporter and photographer.

She covered Tallmadge for a year and has covered Hudson for more than four years. Her beat includes council meetings, new businesses, crime and features. She writes a column under the name, Freeman of the Press. She has written and photographed for several magazines and is a writer for the Spilled Coffee Chronicles[tm] of Invention.

www.ingramcontent.com/pod-product-compliance
Lightning Source LLC
Chambersburg PA
CBHW081139170526
45165CB00008B/2731